From Unicorns & Glow Worms
To Ass Cream & Water Bed Boobs

How To Survive Breast Cancer

By: Erin Raiolo

Acknowledgements & Preface

A good friend once said, "A story isn't a story unless it's shared. Otherwise, it's just a memory. Untold memories don't change lives". The idea behind this book came when I began struggling with the severity and realness of my cancer diagnosis. When I was first diagnosed, I had a hard time believing that I had cancer. I was 33yo, no family history, I was healthy…. How could I possibly have cancer? I started seeing a therapist and she encouraged me to write down my feelings so my mind wasn't full of fleeting thoughts and what ifs. It was then that the book was born. As my journey continued, I began to see a deeper value to the book. Not only was it helping me share my feelings and journey, but I realized my honesty, transparency, and humor could help someone else.

I could not have survived this phase in life without "my people". To my amazing Husband, we've grown together through all of this and our relationship has never been stronger. Thank you for your constant encouragement, listening ear, and holding my hand through it all. You are my rock and best friend. I couldn't have done this without you. To my kids for making me smile and helping me enjoy the little things in life. To my parents for always being by my side and being that unwavering presence I so needed. To my brother, my Chief of Comic Relief, for always lifting me up and making me laugh when I wanted to cry. To my best friends: (Katie, Maggie, & Kate) … your friendships have meant the world to me, especially during this time of my life. To Kristin for the amazing images. You were able to take my crazy ideas and illustrate them! Lastly, to the rest of my family and friends, your encouragement and kind words were appreciated more than I could ever put into words.

As you read this book, none of the information provided in this book should be considered expert medical advice. What I went through may or may not be the same for the next person. I tried to share common reactions to treatment and how to manage. I am also not paid to advertise certain products. Please follow the direction of your medical doctor. The information in this book should be read with a smile. Please also be aware, I am very honest. I tell it how it is. I don't sugar coat but I can be sarcastic. For that, I apologize for the occasional profanity. The tips and suggestions are things I've learned along the way, helped me, and information I want to share with you.

Table of Contents

Chapter 1

The "C" Word: Diagnosis

In January of 2016, I went to the OB/GYN because I felt a painful lump in my left breast. I had noticed it for a couple weeks. To be honest, I've felt lumps and bumps in my breasts for years. I've never really known what to look for as my breasts have always been huge (36DD), fatty, and lumpy. I remember asking one of the docs in the clinic years ago what breast cancer would feel like and I remember she showed me a fake rubber boob with varying sizes and shapes of breast cancer lumps. I remember thinking to myself, breast cancer was supposed to feel like a hard pea fixed to the floor of my boob. So, I just kept the idea in my head that as long as it doesn't hurt and doesn't move, it's OK.

At first I thought it was a pulled muscle or a knot, as I started boxing again after my recent gallbladder removal surgery. So I let it be. Then I got my period, and being a good little nurse I knew it wasn't good to assess the lump a week before or during your period as it can give a false sense of lumpiness and tenderness. So I let it be. A week later I noticed it was still there, it was still painful, so I saw the OB/GYN. Going into the appointment I wasn't too worried as I've always been told I have lumpy bumpy boobs and a history of a benign cyst on my right side. During the appointment, the doctor confirmed the lump, but wasn't concerned because:

1. I was 33yo
2. I had no history of breast or ovarian cancer in the family
3. I had my kids at a young age (under 35yo)
4. I breastfed my kids
5. I didn't take hormone replacement
6. I had a history of lumpy bumpy fatty breasts
7. No nipple discharge
8. I didn't drink in excess, very rare actually. 1-2 drinks/year. (Apparently, alcohol raises the level of estrogen in the body which can lead to increased risk of breast cancer)
9. The lump moved around
10. The lump hurt

For good medicine, we decided it was the right thing to do and work it up. So I was scheduled for a mammogram and ultrasound to be sure it wasn't anything more.

The following week I got to the breast center at 10am for my mammogram. I was told the machine was going to squish and slam your boobs in between these plates and it was going to hurt like H**L. So I got undressed and sat in a lovely, calm room with lavender aromatherapy being misted into the air. The mammo tech was absolutely wonderful. She calmed my nerves and explained what was going to happen. After the first squish I realized it wasn't so bad. There was no indication that my boobs were going to be ripped from my body during the exam. After the mammogram, I went back out into the little calming room with misting lavender and waited for the ultrasound.

The ultrasound tech was also a wonderful human being. She explained everything that she was doing and showed me where the lump was and what it looked like. I noticed she started ultrasounding my armpit and I asked if she was looking in the lymph nodes. I remember wondering if something was wrong but she told me that it's standard to look in the armpit because many breast cancers originate in that area. When she was done she had me wait in the room for the radiologist to come in and talk to me about my results.

When the radiologist came in, she also ultrasounded my breast and arm pit area. She was taking all kinds of measurements and I got suspicious but waited to ask questions. At the end of the ultrasound the radiologist explained that I had not 1 but 3 areas of suspicion. I had the large lump that I could feel, that was 1.5 x 1.5 cm and then there were 2 smaller satellite lesions in the same area approx. 1cm in diameter each. I could not feel the smaller lumps as they were up against my chest wall. She told me because of the suspicion of breast cancer, a biopsy should be done that day.

There it was. THE C WORD. I was in shock. I was devastated. The wind was knocked out of me. I immediately thought it was a death sentence. The world stopped turning... at least I felt like it did. **CANCER?** I'm 33yo, it can't be. I remember walking out of the hospital and crying hysterically. I couldn't breathe. I couldn't catch my breath. I felt like vomiting. I called my mom and she met me at the hospital. I just cried and shook. I couldn't wrap my head around it. My mind kept spinning and thoughts about my family and life kept swirling around my head.

We went back to the hospital two hours later for the biopsy...it wasn't bad. The site was numbed with lidocaine and the radiologist took 4 samples of tissue. When the four samples were done, she placed a marker (aka human tracking device) in the lump for accurate identification and site marking for radiation if it came to that point. Before leaving, the radiologist explained that this was probably cancer and that I should prepare myself for a long journey. She told me to think of 2016 as a book and this was page two in my cancer book/journey.

That night we left for a medical conference, hoping to keep my mind busy and off the idea of cancer. The three hour car ride was filled with mixed emotions. Disbelief, doubt, anxiety, sheer terror. Luckily my Chief of Comic Relief was on hand to blast various 1980s and 1990s movie soundtracks to sing too. Top Gun proved to be just what I needed to keep me from not completely losing my shit.

Let me just say...I HATE waiting. I am a planner and I hate waiting. I was constantly checking my phone the next day during the conference, hoping someone would call with my pathology results. I waited, and waited, and waited.

At 430pm, I got the call. The lovely nurse told me the bad news: **YOU HAVE CANCER**. I was told I have stage 2, grade 2 invasive ductal carcinoma. It was heart wrenching. Everything I was told and everything I learned in nursing school about the characteristics of breast cancer were not true. Not for me. So why did I get breast cancer? I tried to take notes and not cry at the same time but it wasn't super successful. From that moment on, the wheels of treatment and information gathering were put into motion.

Tips For Surviving: Bring someone with you to all of your imaging studies and appointments. Even if you think it could be nothing. I had no idea this pain in my chest was cancer and when I found out, I was alone. It was terrible. From then on I had someone with me at every xray, bloodwork, CT, MRI, doctors visit, everything. Even if they are there just to listen, hold your hand, take notes ... Don't go at this alone.

Chapter 2

Assemble Your Team

The day after I was diagnosed I was assigned to a group of nurses called breast center navigators. They helped me plan all my appointments and imaging studies. Within three days of diagnosis I was meeting with the surgeon and oncologist to plan my course of action and within ten days I had a port and started chemo. Everything moves very fast and it's hard to stop and process everything. Trusting in your team is so important and if you have questions or don't agree, get a second opinion. Be an advocate for yourself.

Primary Doctor: You will continue to have a primary doctor for non-cancer related appointments, and they will stay in the information loop but they won't be a key player on your oncology team.

Obstetrics & Gynecology: Like the primary, this doctor will take a back seat when it comes to treating your cancer. For me, I went to the OB/GYN when I first felt my breast lump and my doctor ordered my mammogram and ultrasound. Once I got my results, I didn't really interact with this doctor again during the course of my treatment. For pre-menopausal women, working with your OB/GYN or fertility doctor may come into play if you need/want to have discussions about having more children/freezing eggs. All women should continue to follow up with their OB/GYN (if you regularly see one) following treatment for annual exams including breast and pelvic exams.

General Surgeon: This is the doctor that preforms the breast tissue removal surgery. Whether you have a lumpectomy or mastectomy, this doctor will be the doctor that removes either the lump or tissue of the breasts. They may also remove lymph nodes if necessary. After the primary surgery, this doctor typically steps back and plastics takes over if reconstruction is something you want/need.

Plastic Surgeon: This doctor will be in charge of all your cosmetic and reconstruction needs. They will work with the general surgeon at the time of surgery to make sure you will have the cosmetic result you're looking for ie: implants, minimizing scars, areola tattoos, nipple reconstruction, etc.

Medical Oncologist: This doctor specializes in the chemotherapy and hormone therapy side of oncology treatment. They are usually the MVP of the team, and while the work with all the other members of the team, they are the ones deciding the course of treatment and who manage your continued oncology journey.

Radiation Oncologist: The radiation oncologist is a doctor that specializes in using radiation to treat different types of cancer. Depending on the stage of breast cancer and lymph node involvement, radiation may or may not be in your cards. Research is showing that even with minimal lymph node involvement, 1-3 nodes, radiation can be helpful when trying to kill any remaining rogue cells.

Family/Friends Support System: Certain family and friends will make their way onto your team. You'll find that people you thought would be by your side will either rise to the occasional or they will fade into the woodwork. Friends you thought you had will stop calling because they don't know what to say. Early on in my diagnosis I found out very quickly who my "people" were. The people I could count on to hold my hair when I puked, carry me upstairs, bring meals, take care of the kids, come to appointments, sit with me at chemo, and help when I needed it.

Other Cancer Warriors: I met some amazing people going through treatment. Family and friends who've survived cancer also speak up and offer to help. Immediately you make friends and these will be "Your Pink Sisters". These are the people who've been there, done that, the ones that truly get it!

Tips For Surviving: Make a cancer command center. I put together a 3 ring binder once test results started coming in. I had the following dividers: The team (all my doctor contact info), imaging results, labs, chemo medication info, chemo visits, surgery info, plastic surgery/reconstruction info, FMLA & STD information for work, etc. I brought that to every visit and updated it after every test. It helps having everything in one place.

Chapter 3

Glow Little Glow Worm, Glimmer Glimmer

Testing Galore

You've just been diagnosed and now your doctors want to send you for "more tests". Below is a list of common tests and what you can anticipate and expect to experience.

Blood Work: It is what it sounds like. Hopefully you have good veins and there is no need for multiple pokes. I had terrible veins so I had bruises up and down my arms. Looked great. If you get a port, the multiple stab wounds will decrease and they will use the port to access your veins and draw labs/give medications.

Mammogram: I heard horror stories about how your boobs will be squished and feel like they're being ripped from your body but that wasn't case for me. The tech will have you sit or stand in front of a specialized xray machine that takes images of your breasts. The tech pulled my boob as much as she could and then the machine squished my boobs in between these plastic plates. There's a chance that because my boobs were so big I only felt a little pulling at my chest and under my armpit and overall pressure. It was not too bad. And the pressure only lasts a couple seconds, just long enough to get an x-ray of your boob. Depending on what they are looking for expect 3-6 images of each boob in different positions. Don't have anxiety about being naked from the waist up, the mammogram tech pulls and pushes and manipulates boobs all day long so throw em out there!

Ultrasound: This is an easy one. You'll lay on a table and be naked from the waist up while the tech/radiologist runs an ultrasound wand over your chest. Once the tech or radiologist is ready you'll expose one breast at a time. They will also ask you to raise and tuck your hand/arm under your head. Aside from a little cold jelly (sometimes it's not heated) and keeping your arm raised and bent over your head it's painless.

Chest/Breast MRI: You will start by getting an IV so they can give you contrast. Contrast is a dye that is used to light up certain body parts for better visualization. Just make sure to tell them if you have shellfish allergies as the contast dye is contains iodine. Once you're ready for the test, you will lay down on a narrow table face down and dangle your boobs into these two camera holes. Then the MRI tech comes and pulls and squishes your boobs into the holes in just a certain way so you can get as much breast and chest tissue onto the camera as possible. Then you place your face into a hole, like on a massage table but without the padding. Lastly you rest your arms over your head and lay flat trying not to move or breathe too deep. For the next 30min you lay there while your boobs dangle in little holes, contrast shoots up your arm and into your chest, and your forehead gets squished. After that, it's over.

CT Scan: This is another easy test. You lay on a small, narrow table and slide inside of a donut as it takes multiple images of one area. It's painless and quick, just a few minutes. CT can also be done with contrast for better visualization. Again, make sure to tell them if you have shellfish allergies as the contast dye is contains iodine.

PET/CT Scan: A PET/CT is like a CT scan but longer. You will need to not eat for 6-8 hours before the exam. This test is longer than a CT scan so plan for 25-30min inside the donut as it scans your entire body looking for cancer that may have spread outside your breasts via blood or lymph nodes. You'll start with an IV and you'll have your blood sugar checked. Blood sugar needs to be within normal range to start the exam, this is part of the reason why you don't eat beforehand. Cancer likes sugar so it's important not to eat so you don't falsely light up. After the IV is placed, you will be given an injection of a radioactive dye. That's when the glow worm process begins. You may also be asked to drink barium; it helps light up your intestines for a clearer picture. After an hour of waiting for the glow juice to circulate you'll be ready for the test. I remember this being the hardest test psychologically. After 24 minutes, my body was scanned from head to toe. As I laid there, I cried softly and stared at the floral mural on the ceiling. All I could do was hope and pray that the glow juice stayed in my breast and no other organs or body parts lit up. Glowing shows areas of cancer. So lighting up like a glow worm in a spot other than your breast is not good.

Tumor/Node Biopsy: This can be ultrasound or CT guided. Ultrasound guided uses ultrasound to show the radiologist where the tumor or node is and where they need to biopsy in real time. A CT does the same thing, however, with CT you lay on a CT table and slide in and out of the CT machine taking images and the biopsy is then guided somewhat blindly. The site is numbed with lidocaine and then they place a metal guide wire into the breast for repeat tissue removal. If it is CT guided, they place the needle, you go into the machine and take a picture. You come out advance the needle a little farther, take a picture, and repeat the process. It's the needle/CT dance. It can be very time consuming to say the least. In either fashion they will use a hollow needle with a gun type device on the end to extract tissue up into the need for removal. Each time they deploy the gun it sounds loud but it doesn't hurt. After samples are removed, the radiologist can use the same guide wire track to place a clip. One of the main reasons for biopsy, aside from diagnosing cancer, was to determine what kind of tumors you have and if they are sensitive to hormones (estrogen, progesterone, HER2). If they are sensitive to hormones, or not, that will help your oncologist guide treatment.

Clip Placement: Clips aka the little human tracking devices. No … not really. They are small metal swirly corkscrew looking things that stay inside of a tumor or node to mark where they have biopsied. They are nice because they stand out on imaging exams and will be easy to locate once you have the tumor or node removed. They can also be used to mark the place of future radiation if that's needed.

Echo: An echo is an ultrasound of your heart. It looks at the size, the valves, the strength of the heartbeat, and how much blood is flowing in and out of the chambers of our heart. If you get chemotherapy (only certain kinds) your doctor may send you for an echo before and afterwards. Certain types of chemo, specifically, Adriamycian can be toxic and hard on the heart. It can actually cause the bottom chambers of the heart to dilate which makes the heart work harder. This is why some chemo meds have a lifetime max dose so it doesn't put extra strain on your heart. The ultrasound is painless but the jelly might be cold!

Tips For Surviving: Again, bring someone with you. Find out when test results will be ready so you have a timeframe to guide your anxiousness. Try to schedule testing on a Monday/Tuesday so you don't have to wait through the weekend. If possible, have someone with you when you get results. This whole process is an emotional rollercoaster and you don't want to ride it alone.

Chapter 4

SMILE...YOU'RE BEAUTIFUL

Take a Moment For You!

You've just got your diagnosis. You're going to start chemo next week. You're going to lose your hair. This is all happening so fast. CRAP. I don't have a moment to breathe.

Take a moment for yourself:

5. Get your hair done. Who cares that it's going to fall out. Try a new color or style. You've got a couple weeks to still look and feel beautiful.

4. Grab an expensive cup of coffee at one of those fru fru coffee shops. Spend $6, sit by the little fireplace, watch the world go by. Relax and have a moment.

3. Go get a massage, pedicure, manicure, etc. In the near future you won't have energy for much. Take the time now to pamper yourself.

2. Take pictures with family. Before my hair fell out, and before cancer took over my life, I had photos taken with my husband and kids. I wanted photos to look at while I was sitting in the chemo chair. I wanted photos with my kids while I looked healthy. I wanted my kids to have photos of me as they envision in their head what "Mom" looks like. TAKE THEM...it's important.

1. Take pictures of YOU. Get pictures done of yourself. I went and got boudoir photos done. I did some in tasteful lingerie to celebrate the boobs one last time. I also brought my Fight For Life shirt and boxing gloves (I boxed for fun/exercise before all this happened) and had fun. The girls at the salon pampered me while I got my hair and makeup done. They told me I looked beautiful for 3 hours, and I believed them.

Tips For Surviving: Throughout this process it's easy to get caught up in the doctor's appointments and the treatments. It's easy to stay inside because you feel like crap or you're afraid to catch a bug. Take time for yourself along the way. You'll feel better to get out of the house even if just for a moment. It's amazing how much energy is packed into a little cup of coffee or a pedicure when you've been sitting at home all day!

Chapter 5

Power Port: Feel The Power

Before chemo begins, the doctor will place a port-a-cath or port. If you have chemo before surgery, this is done right away. If surgery is first they will often place the port during your lumpectomy/mastectomy surgery and then you'll need a second surgery later to remove it. I had chemo first so I had the port placed, had chemo, and then they removed it during my mastectomy.

The port is placed in the chest under the skin opposite of your cancer side. Since I had cancer in my left breast my port was on my right side just below my collar bone. The surgery is "minor". You'll go in for surgery, get an IV, get some antibiotics and fluids and head into the operating room (OR). For most patients, this is done under mild-moderate sedation and not general anesthesia where you would need a tube down your throat to help you breathe.

In the OR they will make a small (approx. 2inch) incision just below the collarbone. The end of the port is a long plastic catheter that goes into a vein under the collarbone which eventually flows into the top right chamber of your heart. In surgery they will make a pocket for the small port and stitch is down to your chest muscle. Your skin is then stitched closed. It's a relatively short procedure, mine took 25min. After surgery, I took pain medication for two days and then switched to ibuprofen. Icing always. Ice, in my opinion, is often forgotten and an excellent healer.

Tips For Surviving: Like I said this is a relatively short procedure with minimal recovery/down time. One VERY IMPORTANT thing to ask is for your port to be a power port. You will have many more imaging exams and if you need a CT scan or MRI where they need to use contrast only a power port will do. If it's not a power port, you'll have to get another IV stick. My doctor gave me a prescription for Emla cream which is a topical numbing medicine. I put a dime size amount over my port an hour before the nurses would access it for procedures, labs, chemo etc. Over time the skin was dulled but it helped with the pokes.

Chapter 6

The Red Devil: Getting Through Chemotherapy Days

Chemo days are LONG days. Plan to spend 5-7 hours/day at chemo. I can't speak to every facility or everyone's chemo regime, but the morning of chemo I would go in at 8:30am and get my port accessed and have labs drawn. At 9:00am I would meet with the oncologist and review how I was feeling, weight, diet, pee/poop status, and she would decide if I got chemo that day based upon my labs. She would also do a physical and we would talk about what worked the last cycle and what didn't. Medication adjustments were made if needed. At 9:30-10:00am I would sit in the infusion chair for my pre-meds. Around 11:00-11:30am I started my actual chemo infusion. Infusions would last anywhere from 1.5-3 hours long. After the infusion they'd let the remaining fluids run in rounding out a 6-7-hour day. All of these times were dependent on the previous task. If labs were bad, no chemo. If labs were good, we were a go for chemo but then the pharmacy had to mix it and send it down. On days one and five (when I started a new medication) I sat out in what I called gen pop or general population (a little jail humor for my deputy husband), instead of a private room so the nurses could monitor me closely in case I had any emergency side effects or reactions.

I remember my first day and I was anxious about accessing my port as it was still tender from surgery so I lathered it up with Emla (numbing) cream. The access went fine so that was good, just a little sore. I started with pre-meds which included Zofran and dexamethasone to help with nausea. I also had a bag of fluids going. After the Zofran was done I got another mediation called Emend which is a long-acting anti-nausea medication. Unfortunately, I had a reaction to that and it had to be stopped. My tongue swelled and got tingly, the roof of my mouth got tingly and my neck and ears got red. No more of that. No need to die before the dang chemo even starts! After that, I was switched to Aloxi.

Once pre-meds were done I got the chemo. We started with the "Red Devil" aka Adriamycin (Doxorubicin). Adriamycin is an anti-cancer medication from the anthracycline antibiotic family. It is very toxic and there is typically a lifetime dose of this medication. It is also dosed based upon weight. It looked like red kool aid. I noticed the nurse pulling back on the syringe as she was giving it to make sure we were still in the vein...if not the chemo can leak into the tissue and kill it. I thought, Ok, pulling back for blood was a good idea. I remember one nurse telling me the red devil med was stealth and because it was red in color it blended in with my circulating blood and the cancer cells never saw it coming. A little guided imagery for ya...whatever helps paint a strong mental picture I guess!

After the red devil they hung the Cytoxan (Cyclophosphamide) for 1.5 hrs and I waited. I ended up sleeping every day. We got a lot of education at the bedside. I learned all about the side effects and the pharmacy of medications I was going to start taking. I learned about good hand washing, foods to eat and ones to avoid, flushing 2x to make sure that my toxic urine and poop got out of the toilet for the next person, washing my clothes separately etc. It was all very overwhelming and scary. Luckily they give you handouts on everything in case you forget. My nurses also recommended bringing someone with for education day, a fresh set of ears.

By the end of the chemo I was felt my stomach growling and felt very dizzy. My mom took me home and I could barely make the car ride. I got so nauseated but I never threw up. I wish I would have, I probably would have felt better. I remember being walked upstairs and lying in bed while my mom rubbed my sore back (a normal side effect is bone pain) and my tummy to help with nausea. I got a cool wash cloth for the excruciating headache. My Dad hit the grocery store for saltines, ginger ale, lemon drops, cough drops, animal crackers, you name it. Anything to get me to eat. All I could do was lay in bed and try not vomit. I took my 4 medications and went to bed that night.

For three days after chemo I would feel extreme fatigue and nausea. I often went into the clinic or ER for fluids because I was dehydrated. I lost 13lbs in the first week and my oncologist told me things could not continue that way. Dang, I thought to myself at one point...OK...high school jeans are gonna fit again! But I knew that wasn't healthy. Around days 5-8 after chemo the nausea subsided but the fatigue remained. I ended up taking medication for chronic constipation and actually developed an anal fissure that would hurt and bleed every time I pooped. Just about the time I was going to go back in for my next cycle (days 12-13), I'd feel "normal" and back to myself all for everything to start all over again. This cycle continued for two months (4 cycles of A/C) while I was on that chemo treatment.

In month 3, I started Taxol (Paclitaxel). I decided to do stacked doses which is three doses in one. This way I only had to do four cycles, once every other week. The other option was 12 cycles, once a week. No thanks, I didn't want to be poked eight extra times even if the side effects were supposed to be less. Nothing changed on infusion days except the pre-meds and the infusion time was longer. According to my oncology nurses, Taxol has a high rate of allergic reaction so they gave me Benadryl, Pepcid, Aloxi (anti-nausea) and Decadron (steroid). All of these medications help fend off allergic reaction and help with nausea. It was also good, per the nurse, that an emergency epinephrine kit along with an intubation kit was standing by in case they needed to put a tube down my throat to secure my airway. OK... good tip!!!! Within a couple minutes of getting the IV Benadryl I was LOOOOOPPPPYYY!! I guess I entertained my husband for a good 30min. He has blackmail videos to prove it. It made me very giggly, irritable, gave me tremors. Eventually, I had to get up and walk around because sitting in the chair one more minute wasn't going to happen. I've had oral Benadryl before but this is apparently a normal reaction for someone who doesn't take a lot of home medications (not used to it) and how fast they gave it. After I walked around the infusion center I fell asleep for 2 hours. As soon as I woke up I asked if I had an allergic reaction. Apparently not if I slept through it.

I had no nausea, no vomiting, and I wasn't as tired. I could eat again. It still tasted metallic but I could eat and drink whatever I wanted. But HOLY bone pain Batman. I again had a cycle. Day one, felt great. Day two, I had burning and redness in my hands. My first infusion I thought I was having an allergic reaction but I learned it was a normal response. On day three, I would wake with extreme muscle and bone pain. The muscle pain felt like I had just done the hardest lower body workout of my life. The bone pain felt like sharp, stabbing twinges in my knees and shins. I also developed neuropathy or numbness in my fingers, toes, heels, and face. With my hands, I felt like I've touched a hot burner and scalded off my fingertips. It was constant throbbing. With my feet, I felt like I'd run a marathon over hot coals and my toes were going to shoot off the end of my feet. The numbness was excruciating at times. Like when I was trying to go to bed. Just touching my heels to the bed was painful and often kept me up at night. Wrapping my feet in a heating pad helped sometimes. I found pain medications wouldn't touch this type of pain. I sat under two heating pads and stretched in the shower. My doctor eventually gave me a medication called Gabapentin for nerve pain, typically given to chronic pain or diabetic patients. According to my oncologist, neuropathy can last years after treatment is completed. The bone and muscle pain lasted four days and then I felt better again. Taxol was a lot more manageable for me than A/C.

The day after each chemotherapy infusion I would go back into the hospital (the 24hr mark) and I'd get a shot called Neulasta (Pegfilgrastim). Neulasta is given to help stimulate the growth or production of healthy white blood cells during what's called the Nadir period. The Nadir period is the low point in between chemo cycles when you have low blood counts. The Nadir period usually starts 7 days after chemo, peaks 10-14 days, and you recover 21-28 days later. This is why some people get chemo every three weeks because it allows the body to make their own white blood cells. If you are on a more accelerated or dose dense schedule like every one or two weeks, they may give you Neulasta. This shot can be very painful (typically given in the back of the arm) as it stings when it's given. Ask for the shot to sit at room temp as this can help with the burning sensation. Just a heads up, this little shot was average $8000. When I first saw the insurance bill I just about dropped to the floor. BUT...in the grand scheme of things...getting an infection and requiring an ER visit or hospitalization would cost a lot more than $8000. That's how the nurses justified it to me anyways!

Throughout chemo, people asked me all the time how things were going and I believe people genuinely care but weren't ready for a real answer. Sometimes I answered honestly and sometimes I put out a blanket "OK" type statement. I don't know if people really understood what it's like to have cancer and go through treatment and that's OK. Thank goodness not everyone has to experience this during their lifetime.

On May 18th, I had my last day of chemo. There were times I thought the day would never come. Days, I questioned the process. At age 33yo, I never imagined I'd be diagnosed with stage 2 breast cancer. It was physically, mentally, and emotionally challenging but was done! It is by the far the hardest thing I've ever had to do and experience. I don't wish it on anyone.

After chemo my family and I went out for BBQ (my favorite food) to celebrate. Everyone wore their cancer sucks shirts...we looked great! It was fun to celebrate being done with this phase! As part of the celebration I did a balloon release!! There was one balloon for each day of chemo. Gone are those balloons and my days of chemo infusions. It felt good to let this phase go.

Tips For Surviving:
In an effort to be more transparent and to help me come to grips with reality, I shared the following on my Caring Bridge Site:

Chemo is not rainbows and unicorns. **IT SUCKS**

1. I went in every other week and got poked with a 1inch needle in my chest. The needle allowed the nurses to draw bloodwork and connect my IV tubing. The positive was that they don't fish around for a vein multiple times.
2. I got 3-4 medications to help prevent or lessen a side effect or allergic reaction before my 3-hour infusion. Those medications in turn had side effects.
3. After the infusion, I was tired for days to a week. I usually felt like myself again 7-8 days after an infusion.
4. I had no energy. Some of this is was due to low hemoglobin and low oxygen but the meds make you tired. I took naps almost every day. I used to box and workout 3-4x/week and during treatment I couldn't walk up a flight of stairs without getting short of breath. It's humbling.
5. Nothing tasted right. Losing taste, or having a metallic taste is normal. Some people even develop mouth sores. Many times things don't sound good so I ate the same meals over and over or I don't eat at all.
6. I didn't sleep. Pain kept me awake at night along with thoughts about how much cancer sucks. I tried different sleeping pills, meditation, rain noises, and some work, some don't. Eventually I fall asleep after being up for hours but I didn't ever feel rested.
7. My mind constantly wandered. Thinking about treatment, thinking about pain, about my kids, my husband, going back to work. Wondering what my future held. Would I live a long life...would I not? Sometimes due to the meds I know what I wanted to say but I couldn't say it. I felt foggy and didn't have clear thoughts. My mind rarely shut off. This fogginess lasted for months after chemo ended.
8. Just when you felt good (a week after infusion), you had to be careful about venturing out into the world because that's when you are at the biggest risk for catching an infection. I was never more aware/afraid of germs in my life. I used the little wipes at the grocery store to wipe the cart, I avoided crowds, avoided sick people the best I could, and I had hand sanitizer everywhere, I didn't venture out of the chemo bubble often. Not even for holiday or family gatherings. Getting an infection with no white

blood cells to fight it is a big problem. Getting the slightest fever would have sent me into the ER and likely incur a 2-5 day hospital stay. It's scary. It's hard to live life like that.

I wasn't trying to complain or be negative. People asked daily how it was going. This was my honest answer. Even when you have a good day, cancer has a tendency to remind you who's in control. I tried to find strength in the people around me and enjoy little moments every day. Some days were good and some were bad. It depended on the day. As you go through treatment it's important to be mindful of your body and the changes it goes through. As I previously stated, the nurses should go through basic education regarding side effects and what to expect. Don't be afraid to call the nurse line with questions or concerns, you don't have to be alone with your worries...that's what they're there for. Better to be safe than sorry when your immune system is compromised!

Chapter 7

Metal Forks & Ass Cream: Side Effects of Chemo & How To Manage

Chemotherapy is toxic. Literally. It is strong shit. It is killing cancer. With that comes side effects (just like any other medication). Each medication has different side effects and your oncologist or oncology nurse should discuss that with you. Below I've listed side effects that I experienced with the medications I had and how I managed at home. Not everyone will have the same medications. Your chemotherapy plan will be specific to you and the kind of tumors you have.

Adriamycian(Doxorubicin) & **Cytoxin**(Cyclophosphamide) – Commonly called A/C. Adriamycin is in a class called anthracyclines and Cytoxan is in the alkylating antineoplastic class (in case you get another medication in these classes but these are pretty standard in terms of chemo care for breast cancer).

1. **Hair Loss:** Your hair will fall out during the course of chemo if you have this combination of medications. ALL of your hair. Head hair, leg hair, arm pit hair, eyelashes, eye brows, pubic hair…. All of it. I guess you could say, not shaving for 4-6 months is one of the perks of chemo??!! Sometime 7-14 days after your first round it will start to fall out… so have a plan. I started my chemo journey with mid-chest length hair. After diagnosis I cut it to shoulder length for a week. Then I got a very short pixie cut with funky colors for a week. The day after my first chemo, I had my husband buzz my head. I did this for two reasons: 1) I had little children and I didn't want them to be freaked out so I did it in stages. 2) I didn't want to be freaked out. I wanted to be in control of SOMETHING. I

wanted the say in when my hair was going to come out. The day of my second infusion (day 14) my head started tingling and that night it started falling out. It never fell out on my pillowcase but I noticed in my hats and definitely in the shower. I remember one morning I started crying because it was really falling out and I called my husband up to the bathroom and he helped me wash my hair that day. I closed my eyes and cried and he rubbed my head to get the hair to fall out. Once it was all said and done I was left with little whispy hairs, it never 100% fell out. As my family continued to tell me that it was growing I finally shaved the rest of it so I could monitor said growth myself. Interestingly enough my eyebrows fell out three weeks after my last infusion (they stayed during chemo) and then started growing back. My hair finally started growing back 6 weeks after my last infusion and I got my first eyebrow wax 11 weeks after my last chemo infusion. I finished chemo in May and by August I was no longer wearing hats, I embraced the buzzed/pixie look. By September (4months later) my hair came back the same color but it was wavy!! Little baby curls and waves everywhere. It's not uncommon for your hair to come back a different color or texture. It's kinda fun actually!

2. **Nausea/Vomiting**: I know it sounds counter-intuitive but eating little bits and drinking actually help. Broths (Better Than Bouillon was delicious), soups, lemon drops, saltine crackers, fruit, Gatorade (stay away from red- I thought I was peeing and pooping blood because of the red dye), carrot cake, whatever it takes…try and eat a little something. A little sugar and salt when you're dehydrated and not eating may help with headaches. It's also important to stick with your scheduled nausea medications. Often times the doctor will put you on a combination of medications so stay on top of those and don't get behind. If the medications aren't working, let your doctor know…I had mine changed 3 times before I found a winning combination. Another thing I was told was that essential oils, specifically ginger and mandarin were supposed to help with nausea. The smells were too strong for me and made me more nauseous.

3. **Fatigue:** This may be the only time you can pull a Rip Van Winkle and sleep for 100 years. SLEEP! There's a reason you're so tired, your body is trying really hard to fight and heal. Chemo will knock you out, if you have a moment to sleep/nap…do it. You'll find there are nights when you don't get the best sleep so if you can nap during the day for a couple hours do it! If you have extra trouble falling asleep or staying asleep, consider asking for a sleeping pill. It helped me on nights where my mind or body wouldn't shut off.

4. **Dizziness:** Some of this can be from the medications, it can also be from dehydration. There were times where I would get dizzy doing nothing, other times it would be from sitting or standing too fast. In the first few days following chemo, I didn't feel safe

driving. It gets better and goes away. Move slowly and stay hydrated! If it continues and you are dehydrated, you many need IV fluids after infusions. I did this all throughout my A/C treatments.

5. **WEIRD Dreams:** When I first started chemo I was having WEIRD dreams. I remember one night I had a dream where I rode a riding lawn mower across state lines to my friend's house. On the way, I saw unicorns running down the side of the road and thought they were so beautiful. About 5 miles from her home the light on the front of my lawn mower went out so I used my cell phone flashlight to guide my way. When I got to her house, everyone was sleeping, as it was 2am, so I crawled into her home via the bathroom window. As I got comfortable to sleep in her tub, I heard splashing in the toilet and saw a baby swimming around in it. WEIRD or what? I don't know if it was the new medications messing with my brain, me being exhausted from the physical and emotional toll, or what, but it was CRAZY.

6. **Runny Nose:** It's not actually a runny nose. Remember when I said your hair falls out? So do the little hairs inside your nose. So your nose will drip constantly. People think you're sick all the time. Snot falls out randomly. It's weird but there's nothing you can do about it. Just have tissues nearby.

7. **Constipation**: Chemo kills rapidly dividing cells aka cancer. It also kills other rapidly dividing cells including: hair, skin, nails, and GI. It kills the digestive cells which can lead to constipation and slow moving bowels. The anti-nausea medications (Zofran, Compazine) can also lead to constipation. Stay ahead of constipation by increasing fiber (or taking a supplement), increasing fluids, and taking a stool softener/motility combination pill (there were days I'd take 4-6 pills just to stay regular and soft). The last thing you want is to strain and get an anal fissure from constipation. Before you know it you'll be rubbing your butt with ass cream and taking pain medication just so it doesn't hurt to poop. Yes, this really happened to me.

8. **Chemo Brain:** Also known as brain fog. Short term memory loss is also common. Stuttering and dyslexic moments are normal. I still get easily distracted and have a hard time focusing. I really got frustrated with this at times because I knew what I wanted to say but the words either didn't come or didn't come out right. It was like my brain and mouth weren't connected. As time passed, and chemo was father and father in the rear view mirror, my brain got more clear. After about 6 months I felt I was able to think more logically and critically about stuff. My words still got mixed up but at least I felt like had control over my brain.

9. **Metal Taste/Loss of Taste/Mouth Sores**: The A/C combination can really affect your mouth and GI tract. Keep your mouth clean (brushed) and hydrated. Swishing with Biotene (without alcohol) helps. If you do develop mouth sores, let your doctor know and they can prescribe magic mouthwash (that's really the name). As for the metallic taste, avoid eating with metal forks and spoons...buy yourself some plastic ones. My sister in law (who's also an onc nurse) made me a gift basket and this was one of the top items – plastic silverware and straws! Don't know why, but it helps. Lastly, if you're struggling to find foods with taste...go for spice and citrus. I always found comfort and satisfaction in a good bowel of chili or fajitas (when I wasn't sick).

10. **Headaches:** I got headaches from being dehydrated, not sleeping, and feeling dizzy. Zofran (anti-nausea medication) also gave me terrible headaches so I stopped taking it and switched to Compazine. You'll have to check with your doctor about what to take for headache relief (Tylenol vs Ibuprofen) but take something so you're not miserable. Peppermint essential oil was also very helpful. I just put a couple drops onto a cotton ball and taped the cotton ball to my shirt.

11. **Acid Reflux:** This was bad. I got so bloated and gassy after eating the smallest meal. I first tried Gas X and ate those like candy. Eventually they stopped working and the burning and reflux got worse. I even started regurgitating my food back up into my mouth. After looking 8 months pregnant and not feeling like my stomach was emptying, my doctor put me on Omeprazole and poof...good bye acid reflux. Within a couple weeks of being done with chemo I was off the Omeprazole.

12. **Loss of Menstruation:** Chemo stops and inhibits ovarian function. So not the most terrible side effect....no period for months! My period stopped after my first infusion. Depending on your age, type of cancer, and chemo this function may return. If you are thinking about having kids, this is a conversation to have with your oncologist/OB/GYN before chemo starts.

13. **Changes In Lab/Blood Work:** Red blood cells (RBCs) contain hemoglobin which takes on oxygen when the RBCs travel to the lungs. As blood flows to the body, RBCs get dropped off into the tissues. RBCs regenerate approximately every 120days. It is not uncommon for your hemoglobin blood values to drop and then slowly come back up after a couple months. Being tired, short of breath, and dizzy are common side effects when your RBCs start to drop. Depending on your facility, your doctor may give you a blood transfusion if your RBCs drop too low. The white blood cells (WBCs) are your body's defense for fighting infections. The normal life cycle of a WBC is 4-30 days depending on the type of cell as there are many different kinds of cells that make up WBCs. As previously stated,

patients cannot have low WBCs for long as they are at risk for serious infection and many times patients are given Neulasta to boost WBC production.

14. **Elevated Liver Enzymes:** This can happen with chemo. This was something that will be monitored closely as your liver is a filter for the rest of your body. After chemo ended, my numbers went back to normal. One thing to be mindful of is taking Tylenol. Tylenol can be harmful to your liver and if your liver is already under stress, Ibuprofen may have to be used but that is something to discuss with your doctor.

Taxol(Paclitacel) Taxol is in the plant alkaloid class and is considered a taxane.

1. **Aches:/Muscle/Bone Pain:** Like I said earlier, this was quite uncomfortable. I never had pain in my arms but I had severe pain and aches in my low back, over the tops of my hips, my upper thighs, and shins. It would ache and throb, sometimes it was sharp and stabbing. I did a lot of resting on the couch, stretching in the shower, and I wrapped heating pads to my legs and low back. This was not something that pain medication helped for me.

2. **Neuropathy(Numbness):** When my hands went numb, that numbness came and went with each infusion. My face was numb but didn't bother me. My feet bugged me to no end. During my chemo infusions and for a couple months after, I had terrible numbness in my heels. After that, it switched to my middle three toes. The pain was excruciating at times and I had a hard time falling asleep as the sheets and covers hurt. At bedtime, I would massage my feet and rub lotion on them before going to bed. On nights I had a really hard time sleeping, I'd wrap my toes in a heating pad. I really had a hard time feeling the bottom of my feet. This lasted for months (8+) post chemo. After the first tumble down our stairs (two broken toes, skinned knee, and bruised ego) and multiple slips in the shower, I bought a non-slip pad for the shower and socks with grippers on the bottom. I know, I know ... you'll feel like you're 80 years old but from the bottom of this nurse's heart...safety first! For me, the best remedy was getting a prescription of Gabapentin from my doctor. I tried treating this at home with things like Ibuprofen and pain medication but it didn't work.

3. **Dry Eyes/Watery Eyes:** My eyes either watered like crazy, looked like I was crying all the time or the were so dry they would stick together. I found there wasn't much to do about the watering eyes, just have a tissue on hand. For the dry eyes I would use warm compresses on my eyes to soften the sticky feeling and goop before putting eye drops in my eyes. If you look for an eye drop, I got a nicer drop than say Visine. I found one called Refresh Liquid Gel that not only hydrated the eyes but it soothed them as well.

Tips For Surviving: In an effort to find the silver lining in this shitty situation, my family and friends helped me create a list of the perks of chemo.

Top 10 Reasons Chemo Rocks:

10- The House has never been so clean and the donated meals have never tasted so good! My family was wonderful and they all came over to steam the carpets and Lysol the furniture. Having cancer, I often thought about the harmful effects of chemicals in the home but at this point I wanted to kill EVERYTHING. The last thing I wanted was an infection so we cleaned from top to bottom.

9- Save money on shampoo/conditioner/ body wash and you never have a bad hair day and you get to wear funky hats made by family and friends.

8- You can eat all the ice cream and ramen noodles you want and no one cares.

7- Easy way to lose 15-20lbs.

6- Every Greys Anatomy re-run is on every afternoon. So are classic movies like Caddyshack, A Few Good Men, Top Gun, Pretty Woman, Tommy Boy, etc. Don't forget the sappy yet predicable cute movies on the Hallmark channel.

5- You don't have to shave for months. (This really sucked when it came back, I felt like an amazon!)

4- You can pull a Rip Van Winkle any time during the day and no one cares.

3- Forgetful? Blame the chemo brain. I often felt like my brain and mouth were not connected and blaming chemo was a safe bet.

2- Call me Daisy...My husband loved driving me around everywhere. We hadn't had that much time together in years.

1- You realize how many good people really care about you and would do anything for you.

Chapter 8

It's Not All Rainbows & Unicorns, It's Ok To Feel

People kept asking me how I felt and I didn't know how to respond. I was grateful for the support and the concern but it can be overwhelming at times. I understood that people around me are going through this too but I HAD CANCER. This will affect ME for the rest of my life.

The entire process was an emotional rollercoaster. Just when I had a good day, I'd have a bad one. Just when I got good news, bad/frustrating news was just around the corner. I remember sitting in the waiting room waiting for my PETCT and I got a call saying the lymph node they biopsied was negative. OK great, my cancer hadn't spread outside the breast. We could skip the PETCT. Not so fast...said the surgeon. They had biopsied the wrong lymph node. What the F**K? Seriously? You are expert medical professionals, how do you biopsy the wrong node? Again...SERIOUSLY?!! Highs and lows, highs and lows.

If I could tell you how I really felt I'd say I was: Angry, sad, scared, I even avoided talking about it. I did a really good job of ignoring it on my good days. There were some days where I felt great and didn't believe I actually had cancer. I didn't identify with it until I looked in the mirror and saw my bald head, dark circles under my eyes, and scars on my chest. I was angry this happened. I was angry at my boob. I feel betrayed. I didn't want to look at or touch my boobs. I didn't want my husband to get anything out of them either. I just wanted them gone. I was sad for what the future held. I didn't want my children without a mother.

I'm still scared of the unknown. I like to be prepared and plan for the future but I had a feeling I wasn't in control and that made me mad and scared. I'm wasn't in the mood to talk about it and when I did I cried. Most days my thoughts jumped from having hope and doing OK to the terrifying thoughts of leaving my family behind. I was terrified that I wouldn't see my children grow up. I was terrified I wouldn't watch my husband walk our daughter down the aisle at her wedding. I found that going to work was a great distraction and that writing down my thoughts was a better way to get the images and words out there.

As I progressed through the process I found it easier to accept the fact that I had cancer. It finally set in. The hard part in my mind was making sense of it all. Cancer affects everything. Your body physically and mentally, your job, your relationships, your outlook on life. EVERYTHING. It turns your life upside down and you have to make peace with it and find a way to go on living.

Tips For Surviving: If you have a support system in place that helps. Find "your person" to talk to and be honest. Support groups can help, so can other breast cancer warriors and survivors. It's important to wait to join a support group until you have all your ducks in a row. This is your cancer journey and you need to make educated decisions about your life. Once your decisions are made, seek out those groups. There are so many amazing people out there who are willing to share their stories. If you don't have support at home, seek support from a therapist or counselor. I found that by talking to a neutral third party (a therapist) I was being heard and my feelings and thoughts weren't judged or dismissed.

Chapter 9

A Letter To The Girls : Saying Goodbye To My Boobs

Before surgery I really struggled with body image issues. Growing up I was very athletic and when I stopped swimming, I gained weight. Weight has been a struggle for close to 10 years but one thing has remained the same... I had great boobs. This was one of the body parts I actually liked when I looked at myself in the mirror. So as surgery approached, my anxiety increased and I started having major emotional struggles. My therapist recommended that I write a letter to my boobs and so I will share that with you.

You girls have been with me for a couple of decades. Back in the day, you were the first sign of womanhood. I still remember buying the first sports bra and white cotton bra with my mom. In high school, I was proud of my C cup boobs. My best friend and I would shop at Hot Topic for the most outrageous bras around. No one ever saw them but it was fun! The purple sequined bra was my favorite!

We have had many memories together. We went skinny dipping in White Bear Lake many times during high school. I still remember the white lace bra I wore the first time my husband got to second base in college. You've been a big reason I love wearing tank tops and dresses, you are my favorite assets. During the times of low self-esteem and lack of positive body image, you girls always looked nice. Lastly, you've given nourishment to my beautiful children. Breastfeeding my babies has been greatest gift and memory you could have given me. I was devastated knowing I had to lose you.

Now you've betrayed me. You've grown cancer. I'm angry at you. I won't even look at or touch you because it reminds me that I'm sick. I won't let my husband touch you. You're forcing me to have surgery and cut out the cancer. I'm scared. Having boobs makes me feel beautiful and sexy. It helps me identify as a woman. Now I'm terrified that I will look like a Frankenstein inspired science project when I'm done with surgery. It's true that after all this is said and done, I'll get perky water bed boobs but it won't be the same. I'll be left with ugly scars that remind me of my cancer. I will be ugly and deformed. Song lyrics have told us that the scars remind us that the past is real and frankly...this is a memory I don't want to remember.

Girls you are killing me and you must die. It will be hard to say goodbye. I know I will cry. I will be hurt and angry. I will be devastated. But... I want to continue living my life and that means you and the cancer must go.

Tips For Surviving: As weird as it seemed, writing my feelings down on paper really helped. I started to think of my boobs as a sign of betrayal and it was easier to say goodbye to such a toxic thing.

Chapter 10

Your Butt Looks Amazing, Where Did You Get Those Genes?

Per my oncologist and general surgeon, I was referred to a genetic counselor. Being that I was diagnosed so young (under age 40), I was an automatic referral. We didn't learn much more than we already knew (I did a lot of research myself) but we learned about the different options for genetic screening. The appointment started out with the counselor making a family tree of which we found no real family history of cancer. We were also told that there is only a 5-10% chance that my cancer has a genetic component. 10-20% can have a familial component meaning there can be a clustering of cancer in the family likely due to genetic and environmental factors. Not likely for me. Lastly 75-80% of breast cancer diagnosis are sporadic meaning there is no determined cause but there may environmental factors with increased age. Being that I was diagnosed young with no family history, the counselor was leaning towards no genetic link and more of a sporadic cause. However, I did decide to do the comprehensive blood testing that looks at 32 different genes. There are over 200 but they can only currently test for 32. In May (5 months after diagnosis), we had already hit our out of pocket max and this was a once in a lifetime test according to insurance so why not go for it. Knowledge is power. Results could be negative giving us no answer to why this happened, however, that would mean no testing for my family and kids which is great. Positive test results would mean testing for my family and kids, give me an answer, and help me make further medical decisions regarding further treatment.

Three weeks later we got the results of the genetic testing. They found no genetic mutations. There is no genetic component to my breast cancer. This was great news for my family and kids. Now no one else has to worry about there being a genetic mutation within the family and thus no one else has an increased risk of developing cancer. The only bummer was that it provided no real concrete answer for me. Unfortunately, I am just 1/200 women that will develop breast cancer in their 30s. There is no scientific reason basis for my breast cancer, it is just a random event in time. My Dad says it best all time, SHIT JUST HAPPENS.

Tips For Surviving: Check with your insurance company to see what is covered before you go. I was told genetics was a once in a lifetime test so I tested everything but certain plans may have restrictions. This is also a HUGE area of research and medicine, take the time to do your research and ask questions!

Chapter 11

Well, You Look Good, You Don't Look Sick:

Top 11 Things NOT To Say To A Cancer Patient

Most people have the best intentions. They hear you're diagnosed with cancer and that's how they look at you and think of you. They feel bad for you. They add you to their prayer list. They've seen someone die of cancer so that's how they think you'll look. They want to know how you're doing and most people genuinely care or feel obligated to ask. In reality, we don't want to repeat our story 100 times and we don't want to talk about cancer 24/7.

11. Please...Let Me Know How I Can Help

Ok.... Sure. I'll just make a list of everything I need help with and mass email it to everyone. RIGHT. Instead offer to bring dinner. Mail a card. Come over and cut the grass or plant some flowers. Send gift cards. Reach out to family to ask how to help. Most people want to help and the response can be overwhelmingly wonderful. But, by saying let me know what I can do to help, it leaves the ball in my court. I don't have time to make a list of things I need help with and it adds pressure to find something for someone else to do and then asking. I was never one to ask for help so this was a hard statement for me. Many times, I was adjusting to a new normal and when I did need help it was last minute. Planning ahead with a meal train or setting aside time on weekends to help with kids is a better plan.

10. I Wonder If You Got Cancer From _____? (being overweight, eating mac and cheese from a box, drinking pop, standing in front of the microwave, having your kids later in life, etc)

The question, "Why me?" is very real. I also caught myself having pity parties and blaming myself after not finding a cause. Many patients have already racked their minds (or are still battling) with the "what ifs" and "whys." I had more questions than answers. I eventually realized that it was OK not to have an answer why because it wasn't going to change my treatment plan and I didn't want one more thing to worry about. Instead, focus on treatment and moving forward.

9. I Know Someone Who Had Cancer…You Should…

Unfortunately, everyone knows somebody who's cancer. There are more and more people being diagnosed with cancer every day. It SUCKS. But…I don't always want to hear about someone else's experience especially if there was a bad outcome. Focus on what I'm going through. Every diagnosis is different. Every person reacts to treatment differently even with the same type of cancer. What I am going through is not the same as the someone you know. Please ask me how I've doing and ask thoughtful questions about my diagnosis and how I'm adjusting.

8. So…Are You Going To Lose Your Hair?

Every treatment is different. Depending on length, course, and medication of chemotherapy will determine if someone will lose their hair. For some patients this is a very difficult topic. Having hair and boobs are features that make us women and can be a sense of femininity. It is a source of beauty and confidence that will be lost in a relatively short period of time. It is dramatic and scary watching your hair fall out in clumps and it's not something we are prepared for or looking forward to. When this happens, retail therapy (new hat/scarf/clothes/makeup) can be helpful and a good distraction!

7. You're So Strong. You Can Fight This. You Can Beat This.

Sometimes a strong image can be helpful. But on days when you don't feel well, this can be a hard thing to live up to. It makes us feel like we can't ever be vulnerable or have a bad day. No one loses their battle with cancer because of a lack of trying. People lose the battle when treatment stops working. Instead, notice times where we've handled situations with grace and style. It allows us to have good and bad days.

6. Everything Is Going To Be Fine

Do you have a crystal ball I don't know about?! No one knows this. Doctor's don't know this. After being diagnosed with cancer, most patients will have a varying amount of fear that they will live with for the rest of their lives. Fear of reoccurrence. Fear it's not fully gone. Fear it will spread. Fear that every bump, lump, pain is cancer returning. By using this kind of blanket statement you're actually dismissing feelings. Offer your support. That way if we get good news, we can celebrate together. If bad news arises, we can cry together.

5. I Think We Should Talk About It

Most of the time we don't want to talk about cancer. It's already consuming life at the moment, let's not talk about it. Talk about anything else. Work, summer plans, gossip, the weather, anything. Anything else to get our minds of cancer for 5 mins. We are still people and want to catch up, gossip, and chat like everyone else. When we're ready to or need to talk we will.

4. Nothing At All

Silence is just as bad as the rest of this list. It's understandable that many people are sad and afraid and don't know what to say. One of the hardest parts for me was losing friends I thought I had because they disappeared and didn't know what to say or do. Don't ignore the situation. I found I developed a lot of resentment towards people when they vanished from my life during treatment and came back when I was done. Offer your time and ear for whatever someone wants to talk about.

3. Everything Happens For A Reason

No. No. No. It does not. I used to believe this statement as the stars seemed to be aligned one morning when my Mom saved my Dad's life by doing CPR when he went into sudden cardiac arrest. But I don't know if I believe it now. For those that look to a higher power and believe this is part of the plan, it's a shitty plan. No one wants to leave this earth before their time. I don't want to leave my husband and kids behind. I don't want to miss their graduations, weddings, grandkids. If there's something I'm supposed to learn from this experience, I hope I learn it then cancer will leave, and I can move on. I want to make the best of my life.

2. This is just a bump in the road. It's just a small setback. It's just one more day/week/month …

I heard this ALL THE TIME. Yes, it is just a bump in the road, today. Tomorrow is another bump (setback/complication). By next week, it will be a freaking obstacle course. Every time I heard, "It's just a"… I wanted to scream. For me, every treatment followed something else. Everything was a domino effect and I tried my hardest to stay on track. For example, I couldn't get surgery until chemo was done. So chemo had to stay on track. Then when surgery came, it got pushed back a week due to surgeon's schedules. Fine… it was just another week. Well really, it was more than that. It meant I was pushing back radiation, reconstruction, and most importantly getting done with treatment and on with my life. Lastly, I got my radiation schedule and I was so excited. If all worked out and I didn't fall behind, I would have ended on my 34th birthday. What an amazing thing to celebrate….no more cancer! Well… I missed one day for the radiation machine being down and I ended that day after my birthday. Yes, bumps in the road happen and there is nothing you can do about it but PLEASE don't blow it off. One less chemo, one less doctor's appointment, or day of radiation. It's a big deal. One day closer to a "normal" life again.

1. Well, You Don't Look Sick. You Look Good.

Gee Thanks. This was hard. At first I thought it was flattering. I took it as a compliment. I actually thought maybe I didn't look so bad. Then I saw reality staring back at me in the mirror. I know people mean well but over time it really annoyed me. I saw a funny quote during treatment that said "I don't look sick and you don't look stupid...looks can be deceiving!" But really...What am I supposed to look like? I know cancer is an invisible disease but I still have real symptoms and feelings. Not every cancer is a death sentence and while most people may have that image in their heads, that's not always the case. It's hard to see the effects chemo has on your immune system, your memory, and all the internal symptoms. Our scars are hidden. You don't see the images or thoughts constantly flooding my brain. While I feel like shit on the inside, I'll just smile for you on the outside so I don't look sick. Going through cancer, chemo, surgery, radiation it is all an emotional and physical rollercoaster. There are good days and bad days. A person doesn't have to look sick to be sick. Instead ask me how I'm feeling and be ready for that answer.

Tips For Surviving: Start a Caring Bridge site and let people know how you're doing. This way, you don't have to talk, you can write how things are going and it reaches the masses. When you're at large gatherings such as holidays, come up with a positive statement and have an action plan to divert the conversation by asking the other person something about them. If push comes to shove write your answer down on a sign and wear it around your neck so you don't have to repeat yourself. No...I'm not kidding...I seriously contemplated this for one of my family holidays.

Chapter 12

My Mama Always Says, "To Cut Is To Cure": Planning For Surgery

The end of June finally arrived (a month after I was done with chemo) and I was ready for my double mastectomy. While they want you to wait a couple weeks to flush the chemo out of your system and you're your counts to come back up so you're safe (not at risk for bleeding or infection), I didn't want to wait any longer than I needed to. Before surgery I had a pre-op physical with my primary doctor and I had basic blood work done to make sure my red and white blood cells and liver function tests where within normal limits post chemo. I met with the general surgeon to review his plan and what to expect during and after surgery. I also met with the plastic surgeon and picked out a new set of boobs! That was a great appointment! While I did have to wait to get my real implants, I picked out a size so they knew how big my expanders would be.

The day of surgery I was nervous and emotional. I took my shower and scrubbed my chest for 8 min (3 extra minutes for good measure) with the required cleanser. I checked into the hospital and my people (my family and dear friend Katie) where there. Looking great in their CANCER SUCKS t-shirts I may add! My husband and I went back right away and I got changed into a gown and little booties. I took the required and mandatory urine pregnancy test. Really? Yup. Hadn't had a period in six months (thank you chemo) but sure...let's do it just to be thorough. The nurse started my IV in my hand...I was saying goodbye to my port as well! I got a liter of fluids along with prophylactic antibiotics.

Before going into the OR, I went over to radiology and they injected blue dye into my areola and around my nipple. Yes, into the nipple. It stung a little like a bee sting but I skipped the local lidocaine anesthetic because I didn't want another poke. I then went back to my pre-op room and waited an hour. In that hour, the dye moved from the nipple area down through the milk ducts and then drained into the lymph nodes. This is how the surgeon knew what lymph nodes to remove during surgery. The thought was that the main nodes (called the sentinel nodes) were the nodes that likely were positive for cancer and thus a vessel to transport cancer to the rest of my body.

The time ticked away and my anxiety level increased. Thank goodness my dear friend was there to calm ALL of our nerves. Anesthesia came in, and explained the process of going to sleep. They also gave me a little medication to calm my nerves and help fend of nausea (which I knew would be a problem from other surgeries). Just before I went back, the general and plastic surgeons came by to check in one last time, sign consent and we were off.

I kissed my husband and hugged my parents and the nurse anesthetist (CRNA) and I walked backed into the room. I laid down on the OR table and quickly asked for the CRNA to give me something to calm me down and put me to sleep. I remember laying there crying as I was very anxious. I kept my cool all the way up until I got into the room and then I lost it. Even though there were staff in the OR with me, I felt oddly alone. As I drifted off to sleep my fantastic nurse held my hand and then I woke up in recovery.

Both surgeries (removal of tissue and reconstruction) took about five hours. The general surgeon started first and removed my breast tissue, nipples, main/sentinel lymph nodes, and as much skin as they could while allowing enough room to stretch and close. He finished his part and the plastic surgeon came in. She started by separating my pectoral muscles from my rib cage in order to create a pocket for my expanders and future implants to sit in. Once the pocket was created, she placed expanders under the muscles, and secured everything by sewing the muscles to the rib cage. By placing the expanders under the muscle it stretched the muscles and the skin to make space for my permanent implants. Some patients get immediate implants but because I was having radiation, I needed the expanders. Per my radiation oncologist, radiating through a permanent implant isn't as successful and he wanted me to have the best outcome so I went with expanders.

Unknown to me, the plastics doctor placed 500cc (about 2 cups) of air into each one my expanders at the time of surgery so I didn't wake up completely flat chested. This was something I was VERY happy about. I knew I was going to be in pain but I was terrified to lose my breasts. They were one body part I really liked about myself. So when I woke up with a little cleavage, life was good!

I got to my room and was told I woke up in recovery, I didn't remember that. My throat was sore from the breathing tube and my chest hurt. It hurt to breathe but I was encouraged to cough and deep breathe so my lungs continued to expand (and so I wouldn't get pneumonia). The great part was that the catheter into my bladder had been removed at the end of surgery and I wasn't nauseated! I was very tired and loopy, to be expected. More entertainment for my family! I took the offered and scheduled pain and nausea medications to stay ahead of the game. I also had constant ice packs on my chest. I had an ace wrap around my chest which added some compression and counteracted the pain. It also kept the incisions and fluid drains secured and safe.

I stayed in the hospital for three days. Being a nurse I was pretty self-sufficient. I stripped my own drains, got up to the bathroom on my own, I walked around with my family. I just needed them to give me my meds and fresh ice packs. On the third day, both surgeons cleared me to go home an off I went. Before leaving the breast center nurses brought me a care package including a stuffed animal to put over my chest under my seatbelt. LIFESAVER. I'm sure I looked ridiculous with a pink puppy dog on my chest but it cushioned all the bumps.

Over the next eight weeks I tried my best to avoid complications. I wasn't always successful but I did my best to avoid another bump in the road and delay radiation.

Tips For Surviving: Stay ahead of your pain. Take your schedule pain and bowel regime medications following surgery and use ice packs! Ice packs are your friend!

Chapter 13

The Pharmacy Called, Your 12 Medications Are Ready:

Recovering From A Mastectomy

Going into surgery I wasn't worried about the physical pain, I was worried about emotional response to such a visible physical change of losing my breasts. After surgery, the pain was excruciating but my new air filled mounds looked great. Complete opposite. You never know how much you use your muscles until they are impaired. Think getting dressed, pick up your kids, putting away a gallon of milk, washing your face, wiping after going to the bathroom. Doing basic tasks around the house were difficult. Thank goodness for family who cooked and neighbors who mowed our grass. I constantly had ice packs on my sides and on top of my chest to help with the sharp pains and constant pressure against my ribs. Pain was also very tight in my back as the muscles from my chest were pulled forward.

You're not allowed to reach over your head, lift more than 10lbs, put on your own seat belt. Little bumps in the road were painful so I shoved my little pink stuffed dog under the seat belt to cushion my chest. I took myself off narcotics as I didn't want to deal the side effects and I quickly realized being a hero helps no one. Being in pain delays healing so my husband encouraged me to begin taking pain medication again, at least to help take the edge off. My lovely husband made a spreadsheet of all my medications (there were 12 of them) and made sure I took them on time so I could stay ahead of the pain. I couldn't lay flat so my husband or kids slept downstairs on the couch while I sleep in the recliner. My husband and four-year-old son became wonderful nurses!

Every day got a little better. The pain lessened. Getting rest allowed for more activity. I took me about four weeks to feel safe driving. I also had to wait until I was off pain medications and the drains were removed. It was about five to six weeks until I felt completely self-sufficient.

Tips For Surviving: Recovery was trying. It's got better every day but sitting at home in the recliner was not all rainbows and unicorns ... even on pain medication! It was just me and my trusty ice packs.

Chapter 14

Water Bed Boobs: Breasts Under Construction

I had a rough recovery process following surgery. It was all one big rollercoaster of pain, emotions, and ups and downs from complications and setbacks.

Week 1: A week after surgery we met with the oncologist, general surgeon, and plastics for post-op follow ups to go see how I was recovering and to discuss my final pathology report and treatment going forward. My oncologist was very happy with my response to chemo. My kind of tumors liked hormones (estrogen and progesterone) and I was told it is very uncommon for there to be a complete response to chemo alone. Usually there also needs to be hormone suppression (aka Tamoxifen- an oral pill that I will take for 5+yrs) to fully get rid of the tumors. After surgery, we found that I had a COMPLETE response, chemo killed my tumors and lymph nodes as there was no evidence of cancer on final pathology from surgery. My general surgeon was confident he removed all the breast tissue and he was happy with the incisions and how I was healing. GREAT NEWS!

Week 2: Plastics was rough. I was told I'd have my drains in for 10 days. At the 5-day mark, I had to go into the ER as they were full of clots and not draining as much as they should. If the drain was opened or replaced there was a risk of infection so the ER doc let them be and I saw the plastics doc two days later. On day 7, plastics changed out the bulb of the drain and left them in to continue draining. I was sad. I wanted them out. I wanted to shower and drive. They said come back in three days for reassessment. We went back on Monday and only the left side was removed as the right was still draining. Again sad and disappointing. They said come back Friday and we will reassess. Friday of that week I got my right drain out yay!!

Week 3: After close to 3 weeks I was finally able to shower, move my arms more freely and drive! Once the drain was removed, they started the removal of air and addition of saline to fill my expanders. The right side was no problem. They found the little port in the expander, removed the air placed during surgery, and added 500cc of saline. Literally I saw my baby boob growing. Weird but cool! They tried the same process on the left side 3x but struggled. They were confident they found the port but they had resistance each time they tried putting the needle into my chest. On the 3rd attempt they were able to remove the air but they had resistance putting fluid in. During the procedure I had extreme pain under my collarbone and arm pit. I knew something was wrong. So they stopped and aborted for the day. I left the office with half an inflated boob on the right side and a deflated boob on the left side. Can we say Frankenstein boobs? I was a walking, talking, freak show with one boob. Ridiculous!

At that time, they were concerned about the left side for 3 reasons: Best case... the port is in an odd position and they need to make adjustments on how to access it. Worse case(s), the expander was popped during the 3 needle pokes and is leaking into my chest instead of the expander or the expander has shifted in the last 3 weeks and the port is not in a place where they will be able to access. In both of the "worst case scenarios" I will need to go back into surgery to either have it replaced or moved back towards the front of my chest. NOT what I wanted to hear. So in order to determine what was wrong I had an ultrasound to see if there was fluid around my expander, to see if the expander was punctured, or to see if the expander was migrated or flipped into a non-accessible position.

All of these things are common and were minor roadblocks. But as enough built up they began to be frustrating. The simple fact of wanting to take a shower became a big deal. I didn't want another surgery regardless of how "minor" it was. As the process was delayed due to these "minor" roadblocks, it delayed me starting radiology, it delayed me going back to work, it delayed me moving on with my life and saying goodbye to cancer. It's not just one set back it's a bunch of little ones.

The next few days were frustrating. I was sore from the poking, prodding, and the expansion process. The expansion never hurt but I always had ibuprofen on standby. Within a couple hours of expansion, I was SORE. My little nurse (my son) kept the ice packs coming my way and we watched lots of movies! I rested that weekend and hoped and prayed my ultrasound went well on Monday.

Week 4: Monday came and the ultrasound was a DISASTER. The radiologist was sad to tell me he couldn't find anything. Excuse me "head of the radiology department" …. I'm sorry… what? Yup…couldn't find the port, they thought the expander collapsed and the fluid they placed the Friday before was around the expander in my arm pit. SUPER. I was devastated leaving radiology. I thought for sure I was going to need surgery to have the expander replaced. On Tuesday, I saw the plastic surgeon and she was convinced the expander was fine. She looked at imaging herself and saw the port was tipped. Sounded like my plastic surgeon needed a double board certification in radiology. So with double the length (2inch) needle and me laying slightly elevated vs flat, she got it and put 250cc in my left expander. Half of the right size amount but at least it has some volume!

After the expansion, I showed her my incisions. The black, flaking, leaking nasty incisions. I showed the doc and she was equally concerned. I had black dead tissue where the three places of skin came together called the T zone on both breasts. Dead tissue, I learned, is not uncommon because all the nerves and blood supply are cut in surgery and the remaining skin doesn't always get good blood flow. She said we could 1 of 3 things. 1) let it heal on its own and scab over the next 2-3 weeks and hope it doesn't get infected. 2) clean it out in clinic with no pain meds. Um...No Thank you 3) go into surgery now and clean it out and re-stitch the incision so it can heal.

We chose option 3.

Friday of the 4th week I had my second surgery. My Dad was my nurse for the day since the rest of the family was on vacation. For whatever reason I was more worried about that procedure than my mastectomy. I was a nervous nelly! I made sure to ask for extra nausea and anxiety medication that day. The plastics doc told me after the procedure that on my left side the tissue was superficial and close to the surface so she cleaned it out and restitched it. On the right side, the tissue was dead and there was evidence of infection all the way down to the inside pocket so she cut the incision wider and washed the pocket out with an antibiotic solution. She also had to drain extra fluid out that had built up around my expander called a seroma. Likely my drain was taken out too early but seromas were a common complication.

Week 5: While I was waiting for my incisions to heal they wouldn't put any more fluid in to expand my expanders. I was in limbo. I had to wait and be patient.... not so great at that. Again...a little bump in the road but it meant no radiation, no work, had to put my life on hold some more. Everything was a domino effect and I just wanted to move the process along.

Week 6: After the second surgery I continued to have seromas (fluid pockets). When your body undergoes surgery or there is some sort of trauma, fluid weeps out of your vessels and collects to help heal that area. I eventually had 5 seromas drained from my right side and two from my left. The amount of fluid ranged from ½ - 1 full cup of fluid each time. If left alone they would have eventually been absorbed by the body (30cc/day) but the extra fluid was causing extra pressure and that was uncomfortable. In order to remove the fluid, the doctors used ultrasound to make sure they don't pop the expander and then a needle to remove the fluid. The doctors never gave me a good reason as to why I kept accumulating fluid, it was just something my body did.

Week 7: I finally had 3 great plastics appointments and I had saline fills each time. Each fill stretched the muscle and the skin so it can accommodate a regular implant in about 8 months (have to finish radiation first and then wait 6months). When I got a fill, they used a magnet to locate the port of the expander and then they shoved a needle thru the skin and muscle into the port. Once in the port they added fluid (50-100cc) to fill the expander. Hurt like a ... I usually found myself cutting off circulation to my husband's hand or ripping the cushioning off the exam bed. The muscles in my back hurt as the chest muscles were pulled forward. I finally got up to 800cc and I had a final ultrasound to confirm all was well and I was ready to start radiation.

Approximately two months after surgery I was able to move around freely and on my own. In an attempt to increase strength and flexibility my plastic surgeon sent me to physical therapy. I went for 6 sessions and got frustrated because I wanted to go back to working out and boxing, not lifting soup cans over my head. Did I age overnight? I didn't think I was 80 years old. Frustrating. I took it upon myself to take the basics and add weight and intensity. I had a life to get back to.

As time went on it was easier to do normal everyday activities. Sleeping was the only thing that was uncomfortable. It is still uncomfortable at times. Gone are the days of sleeping on my stomach. It was really hard sleeping with bricks (expanders) on my chest. I found that sleeping on my side with a body pillow wedged in between my boobs helped. Wrapping my arms around the body pillow took the pressure off my chest and I was able to sleep. I'm sure my husband would have preferred to be my pillow but he wasn't as fluffy.

Tips For Surviving: Ice and ibuprofen...were my best friends. Make sure to have plenty on hand. I found wonderful little ice packs in the baby aisle at target and they were the perfect size to shove into my ace wrap and sports bra while I had the drains in. Don't get down when things don't go as planned. Looking back, I wish I wouldn't have set my expectations so high. I wish I would have relaxed more and went along with the process so I didn't get so disappointed when things didn't go as planned. Take a moment for yourself, breathe, you will get through it. It will be OK!

Chapter 15

Who Is That Shell Of A Person Looking Back At Me In The Mirror?

Cancer changes everything. It changes you physically, emotionally, spiritually, psychologically. It affects relationships. <u>Everything.</u> Emotionally I was a wreck at times. I would be OK one minute and I'd start crying at the drop of the hat. I was anxious and maybe even depressed. For months I couldn't look at myself in the mirror. I think it was a protective mechanism so I didn't have to deal with the fact that I had cancer. If I didn't see it, then it wasn't true. I didn't want to deal with it.

Then one day I looked at myself. I really looked. When I did, I saw a shell of my former self. I was a fresh baldy, thank you chemo. Gone were my breasts and I was left with perky round bricks with a giant scar down the middle of each one. Can we say Frankenboob? My face was round from the steroids and I looked kind of bloated. Just under my right collarbone was my port scar. I could cover my boob scars with a shirt but that one remains. It was a constant reminder of the journey and for a long time, it was something I wanted to forget. Over time I became proud of my scars and all that I endured but that took a good year.

Cancer has a way of humbling you. Before I was diagnosed, I was healthy. Not high school jeans skinny, but healthy. I was boxing twice a week and working out. Eating right. I had four jobs (ER nurse, photography, RN instructor, Mom). I was constantly on the move. Then I got diagnosed and within six hours of my first chemo treatment I was throwing up on the bathroom floor and my Dad was carrying me upstairs to bed. I spent days on the couch not moving, not thinking, just existing watching the world go by. Watching my kids and husband play outside while I tried not to puke or move. It made me smile to hear their laughs but I almost inevitably ended up crying because I was now an absent Mom. Once I was done with treatment I half expected myself to jump back into life as it was before treatment. WRONG. It took a good three months before I felt "normal". Just when you think you're moving on, there's always a reminder. Every time you look in the mirror; the scars look back at you. Humbling.

Tips For Surviving: Talk to a neutral party. Don't harbor your feelings. Set realistic expectations for yourself and create a new normal with new goals. It's not that you won't achieve them, it may just take a little longer. If the scars bug you as much as they bugged me.... cover them! A year after my port surgery I got a tattoo to cover it. The next big piece will be a full chest tattoo to cover my mastectomy scars.

Chapter 16

The Birds & The Bees

Let's talk about the birds and the bees. It is not uncommon to completely lose your libido along with your hair. Cancer treatments can dry things up, you're exhausted, and who wants to have painful "relations"? Then you cut off your boobs and all the nerve endings to that once erogenous and pleasure area of your body are gone. The old methods of arousal and excitement don't work. Sure, your significant other may be giving you the googly eyes but the last thing you want to do is get naked. Then they'd see you. You can't hide! Who wants to have sex if it hurts?

Eventually it gets better. The fatigue lessens and your body begins to bounce back. When your scars heal take yourself shopping for a new bra (no underwire for 6 months) and panties set. Something you feel sexy in. If you feel sexy, you'll have the confidence to show it.

Tips For Surviving: Invest in some lube. It helps with dryness so sex doesn't hurt. Find new ways to have intimate fun with your partner. If it continues, talk to your doctor.

Chapter 17

Endocrine Therapy: Emotions Are Like...A Roller Coaster Baby!

Tamoxifen

My cancer was estrogen and progesterone positive which means those hormones fed my tumors and made them grow. After all was said and done (chemo, surgery, radiation) I still needed medication to suppress my hormones. My oncologist started me on Tamoxifen. The plan was 5 years of therapy but new research is showing success with 10. When I first started I was miserable but eventually, as will everything else, it got better. Side effects included:

1. **Nausea:** I was miserable and had no appetite. Most people gain weight on Tamo but I lost weight because it made me so nauseous. I ended up taking it before bed and then I didn't notice the nausea anymore.

2. **Night Sweats:** I would wake up multiple times per night sweaty and lying in a pool of wet bed sheets. I did a lot of laundry! I never really found a trick to making this better. It just got better over time.

3. **Hot Flashes:** Hello Menopause! Being Tamo is a hormone suppression med, it basically puts you into menopause, if you're not already of that age. I was 33yo at the time of diagnosis. Hot flashes came at the worst times. Working as an ER nurse, I remember we got a trauma one day and we keep our trauma rooms at 70 degrees. With 15+ bodies in there the temp was rising more. I had severe hot flashes, got lightheaded, and almost passed out. For the most part, I felt my hot flashes around the time when I needed to take my next dose and throughout the night. To help not have as many during the day, I took it at bedtime.

4. **Emotional Rollercoaster:** I was a Mega B**CH! One minute I was happy, the next sad and crying, the next pissed off at the world. It was bad for a few weeks. My oncologist recommended I start taking Effexor to help with hot flashes and the emotional side of things. I never took it. I had been on so many medications that I didn't want to take one more thing. I ended up going to see my therapist every week to talk through some of my irrational thoughts and she helped me with some re-directional behaviors for when I would catch myself getting whacky.

5. **Hip/Joint Pain:** Joint pain is a common complaint. My hips would be so stiff in the morning when I got up. Moving around and stretching along with a heating pad helped.

6. **Muscle Cramps/Weakness:** Out of all the symptoms, muscle cramps were the worst. I would get the most excruciating cramps in my calves, mostly in the middle of the night. I started taking ibuprofen but that didn't help. I read online, on a support group site, that people had success with taking magnesium and calcium supplements. I tried magnesium and within two days my muscle cramps had gone away completely.

Tips For Surviving: You have to determine if the side effects are worth your sanity, happiness, and health. I was miserable for a while but I kept telling myself…. "Whatever I need to do so this freaking cancer doesn't return". There are very serious side effects associated with this med so talk to your doctor about managing it all.

Chapter 18

Of Course I'm Glowing...I Just Had Radiation

Radiation started with a planning and consultation appointment. I remember asking the doctor, since I had a complete pathological response to chemotherapy (chemo killed everything) and since I had a double mastectomy...was radiation needed? He said yes. He explained that, we did chemo to kill the tumors, we did surgery to remove the tumors, and lastly we do radiation to treat my chest, remaining lymph nodes, and underarm. The goal of radiation was to treat any remaining rogue cells and the remaining lymph nodes so if there are cancer cells remaining we would kill them and so they didn't spread. OK FINE, I'll do radiation.

After we chatted a bit, I went across the hall and had a CT scan done to look at the organs in my chest. Before the CT images were taken they made a cast of my upper body so that when I did go in for radiation, I would lay in the same position every time. The radiation doctor, a medical physicist, and a medical dosimetrist (radiation dose doctor) then decided how they would radiate my chest based upon the images. They were able to decide the direction of the radiation beams based upon the images and they also used the pictures to make sure they avoided the areas of my chest they didn't want to radiate (heart/lungs). After all three doctors decided on a course of action, the dosi doc tattooed 6 small black dots on my chest so they know exactly where to send the 6 areas of radiation. The one small, tiny problem... my right water bed boob was too big so I had to go into plastics the next day and have 100cc removed.

I had 6 weeks of radiation. 28 days. I went every day (except weekends). I would go in every day, change into a gown that covered just my chest, got to leave my pants on. I'd go into the radiation room and the two radiation girls would get me set up in just the right positon on the table according to my tattoo markings. After that they would take an xray of my collarbone and upper arm bone to make sure they were shielded from radiation. If they weren't shielded they could actually be fractured. Once I was situated and they were confident they were radiating only what they wanted to, radiation to my tattoos took 30 seconds each so a total of 3 minutes. It took longer to get set up and in the exact same position every day then it did to radiate.

I was told common side effects of breast radiation included: sun burn to the treated area, skin/tissue damage and peeling, fatigue, and damage to the top of my left lung (won't be noticeable). Not so common: pneumonia, broken ribs, nausea/vomiting, radiation to the heart, further lung damage, blisters and open wounds. As radiation started accumulating (around treatment 10) the fatigue really hit me. I started taking naps in the afternoon. Over time I surrendered and just took care of myself instead of pushing ahead. While it wasn't common I also developed nausea post treatment and so I took nausea medication as needed. Week 5, I had a noticeable sunburn on my neck and around my collar bone. The radiation doc said this was normal because the skin is thinner, it's been previously exposed to sun, and there's little to no fat underneath. Week 6, I had a very distinct line of redness and pain under my treated armpit. It pulled and was painful. I tried my best to keep the skin hydrated and I added lavender to my lotion to help it heal.

In order to keep my skin moist and healthy I did the following skin care regime every day during treatment (including my weekends off) and up to two weeks after (you still burn that long).

- In the morning I'd shower (sometimes every 2 days) and if my skin needed moisture I'd put aquaphor on.
- After radiation, while I was in the dressing room, I'd put a combination of Calendula cream and Lanolin oil on my skin.
- Before going to bed I'd rub on the prescribed Mometasone cream. Mometasone cream (a steroid) has been in numerous research articles in the last 5 years and is associated with better skin outcomes during breast radiation.
- Once I developed a sunburn (around week 4) I increased my aquaphor to daily and I added lavender essential oil to that to help my skin heal and repair. The goal was to not have my skin open during or after treatment.
- After treatment my sunburn opened and I was concerned about infection so my doctor prescribed Silvidine cream which is an antibacterial cream. I put this on 3x/day and covered it with a breathable gauze bandage so the burn was protected and my clothes didn't get ruined.
- I also found an aloe cooling pad, made by Lindi, on amazon. It was HEAVEN on my burning skin. It was a little pricey but it was reusable and by keeping in the fridge it not only stayed cool but extended its life.

On my last day of treatment my skin peeled. Go figure. When it peeled, it STUNG...BAD. I was told that because you continue to burn for up to two weeks after radiation is done, it would get worse before it got better. The area by collarbone peeled first and continued to redden. As it started healing it began really itching. I tried my best to keep it hydrated. Four days after radiation was done, the skin under my arm pit and under my boob by my incision also peeled. The incision also hurt quite a bit and I put gauze over it so it wouldn't rub on my sports bra and shirts. The burn continued to worsen for 10 days after treatment and then I noticed new skin being formed and the redness turned to a dark pink. It took a good 2-3 weeks for everything to completely heal.

Tips For Surviving: If you know you're going to be tired, get important tasks done in the morning so you can rest in the afternoon. Your body is working really hard to repair the damaged skin and cells and this takes energy. Take care of yourself and let your body recover.

Chapter 19

If a sniffle = a cold, does a headache = mets to my brain?

As life went on and I got back into my routine, moments of anxiety creeped back into my life. I was nervous to get sick so I continued to have hand sanitizer everywhere I went. I washed my hands raw at work, to the point they began peeling. The fact is, your immune system takes a big hit with chemo and treatment so my concerns were valid, however, it shouldn't have taken over my life. Easy to say now, but in the moment it's terrifying.

A month after I finished radiation I got sick. I was scared shitless. Luckily my Mom and Brother (chief of comic relief) came to my rescue and kept me entertained as I sat in radiology for about the 20th time this year. I started feeling a burning in my chest and was short of breath. I knew I was high risk for blood clots being on the Tamoxifen so my oncologist wanted me to have a CT scan right away. I went in for labs, which were normal, and a CT scan. The good news was that I did not have a blood clot but the CT scan did show all the damage from treatment. My heart was enlarged from my heart having to work harder post chemo than normal and there was evidence of damage to my lungs. I also had pneumonia. I was told it wasn't common to get pneumonia post radiation (if I did it would have been around the 6month post treatment mark) but when had I ever followed the "statistics"? In retrospect, I should have bought a lottery ticket every time someone told me the statistics of my cancer.

When they ask about fear it's not the bees, spiders, or snakes kind. Fear is there and it is real. Reality check. It took a few talk therapy sessions to understand that while the concerns are valid they cannot consume you. I found that if I kept myself busy I didn't think about cancer as much. It was when I was alone or with my kids that I thought about it. I often thought about how much it could have taken away from me and how much I vowed to never let that happen. I often wondered if cancer was still brewing in my body and growing although I wouldn't know for years.

Looking back at my fears over time, rationally I knew, a sniffle didn't mean the flu. It was probably just a cold. Pain in the chest may just be pain, indigestion, or a pulled muscle. It wasn't a heart attack or lung cancer. A headache didn't mean the cancer is back and has mets to my brain. The thoughts crossed my mind but it wasn't true. Fear will always be there but you have to trust in the process and your doctors. You've put yourself through H**L, it's normal to feel that way.

I met some pretty amazing cancer warriors during my journey and it was remarkable to see how many of them were success stories and living beyond cancer. It's always positive to hear survivor stories. I also met some truly wonderful people who lost their battle and passed away from cancer. People, I think about often and miss. I trade places with them in my mind sometimes and throw myself a pity party but my amazing family brings me back to reality. It's hard not to think about all the what ifs but I have to continually remind myself ...will it change anything? Will it change treatment? Will it change the plan? Will it change my life? Those are usually answered with a big fat NO and reality comes back into view.

After treatment people would say to me, "Wow, you look great, aren't you so happy that you have that whole cancer thing behind you"? I remember thinking to myself, you really have no clue about this "whole cancer thing". But, I would smile and say yes to be polite. Honestly, it never truly goes away. Cancer is always there. Maybe not physically but in your mind. There's not a day that goes by that I don't think about it. Follow up appointments will continue and you will have a lifelong relationship with your oncologist. It's not like strep where you take a pill and you're cured in 10 days. I wish! But really...cancer was a part of our life but it's not what defines us. We have the ability to write the next chapter.

Tips For Surviving: I know it's hard but try not to let your thoughts consume you. Cancer is everywhere. Just when you think you've gotten away, you hear of someone else being diagnosed. Stay positive, stay the course. Live your own life.

Chapter 20

A New Normal: Life After Cancer

You just survived cancer...now what? Over the course of my diagnosis and treatment year I had over 140 appointments, almost one every three days. You get used to seeing your doctors and nurses. You feel safer because you're constantly being monitored and checked. In a strange way, I felt sad leaving on my last day of radiation because I'd seen my nurses and doctor every day. Afterwards there was a small void. These people have come into your life, hopefully for a short time, and you get to know them and they get to know you. You take an interest in one another. It's kinda of like a sick dysfunctional family. But then treatment stops and you wonder...what do I do now?

I remember laying on the table during my last radiation treatment and I started crying. I couldn't believe that it was over. All of it. Nine months of tests, procedures, and treatments. It was all done. Like always, my husband was by my side and he comforted me. He understood my tears, knew they were happy in nature but also knew me very emotional response.

Again, I asked myself, what do I do now? **You LIVE. That's what you do.**

Yes, you will have a lifelong relationship with oncology but gone are the weekly appointments. Now you will be on the 3 or 6 or 12-month schedule. Cancer is scary and the fear of reoccurrence is real. Follow-up appointments can create a tremendous amount of anxiety but you have to trust that all these months of treatment were worth it and the path you took, killed that darn cancer.

Easier said than done, I know. I struggled for a couple weeks not knowing what to do or how to feel. I wished the doctors and nurses would have stayed in touch a little more in the transition period instead of cutting me lose and wishing me luck. "Don't let the door hit you're A** on the way out". I felt lost at times. I was so used to having someone there to reassure me, I began questioning every little thing. It's like my rational and more specifically my nurse brain stopped working. It got easier as days passed. I remember being at work, back with friends, back in a routine and I caught my reflection in the ambulance bay and I saw my short hair. I couldn't escape the fact that this all happened. I felt like, for a while anyway, that I was constantly taking care of fellow oncology patients. The reminder was everywhere.

For a while there will be a transition period where you have short hair and people still stop and stare wondering why you decided to have a meltdown and shave your head. You could say you went to a beauty school on the first week of class and the haircut was so bad this was the only option. You could say your kid put gum in your hair and you had to cut it short. You could also tell the truth but be ready for a slew of questioning, even from strangers. When the time is right, take off the scarfs and wraps and let the wind whip through your baby hairs.

Go grocery shopping again and touch the cart handle, sanitize later of course! Venture outside and get a little sun on your cheeks. Take yourself on vacation, you deserve it.

Go on living your life. Don't let cancer take away one more day. Have a pity party but set a time limit. It's not uncommon to have a flood of emotions as you begin to mentally process everything but don't let it diminish all of the hard work you accomplished on your journey. Celebrate life's little moments and find the joy in every day. One of the best quotes I found said, "Dear Cancer, Thank you for making me stop and listen and remember what's truly important. You can go now".

Tips For Surviving: Get your LIFE back. Go back to work, play with your kids, go on dates with your significant other, take time for yourself. Your body has been through H**L and back, give yourself a break and be kind to yourself. Find the meaning of life.

Chapter 21

Pay It Forward: Help The Next One In Line

Going through this process I've learned cancer is everywhere. Someone new is being diagnosed all the time. As that happened, I found it extremely powerful and moving to pay it forward. If I did hear of another woman being diagnosed, I'd send a care package of things I found helpful during my journey. I also completed random acts of kindness (brought flowers to a neighbor, paid for someone's coffee, bought lunch for a police officer, brought cookies to our local heroes, etc) and that felt just as good. You'll find that you'll struggle with how to thank everyone. The ones that are with you day in and day out don't do it for the thanks but I always wanted to try.

Towards the end of radiation, I started walking in an attempt to get ready to go back to work. Being an ER nurse it wasn't uncommon for me to walk 3-5miles in an 8hr shift. I knew that while I was resting and napping to heal during treatment, my muscles were getting weaker and weaker. I had to get back on my feet and get moving.

One morning I was walking and the sun was out, there was a nice cool breeze, and I was listening to the song "Humble & Kind" by Tim McGraw. The last line hit me. Hit me hard, to the point of tears rolling down my face. In the song he says: *"When you get where you're going don't forget to turn back around. Help the next one line. Always stay humble and kind."* The next one in line...the next cancer warrior.

I decided in that moment to finish this book. I wrote this book for me. To cope and as a way to share my feelings and be transparent with everyone around me. I wanted to raise awareness and educate. Over time I thought ... maybe my story and humor could help someone else. I wanted to share my experience and hopefully... help the next one in line. Even if it just made someone laugh, that would be enough.

www.ingramcontent.com/pod-product-compliance
Lightning Source LLC
Chambersburg PA
CBHW072045190526
45165CB00018B/1686